Frontispiece

Advanced Lunar Base

In this panorama of an advanced lunar base, the main habitation modules in the background to the right are shown being covered by lunar soil for radiation protection. The modules on the far right are reactors in which lunar soil is being processed to provide oxygen. Each reactor is heated by a solar mirror. The vehicle near them is collecting liquid oxygen from the reactor complex and will transport it to the launch pad in the background, where a tanker is just lifting off. The mining pits are shown just behind the foreground figure on the left. The geologists in the foreground are looking for richer ores to mine. Artist: Dennis Davidson

Space Resources

Overview

Editors

Mary Fae McKay, David S. McKay, and Michael B. Duke

Lyndon B. Johnson Space Center Houston, Texas

National Aeronautics and Space Administration NASA

Scientific and Technical Information Program Washington. DC 1992

Technical papers derived from a NASA-ASEE summer study held at the California Space Institute in 1984.

Preface

Space resources must be used to support life on the Moon and exploration of Mars. Just as the pioneers applied the tools they brought with them to resources they found along the way rather than trying to haul all their needs over a long supply line, so too must space travelers apply their high technology tools to local resources.

The pioneers refilled their water barrels at each river they forded; moonbase inhabitants may use chemical reactors to combine hydrogen brought from Earth with oxygen found in lunar soil to make their water. The pioneers sought temporary shelter under trees or in the lee of a cliff and built sod houses as their first homes on the new land; settlers of the Moon may seek out lava tubes for their shelter or cover space station modules with lunar regolith for radiation protection. The pioneers moved further west from their first settlements, using wagons they had built from local wood and pack animals they had raised; space explorers may use propellant made at a lunar base to take them on to Mars.

The concept for this report was developed at a NASA-sponsored summer study in 1984. The program was held on the Scripps campus of the University of California at San Diego (UCSD), under the auspices of the American Society for Engineering Education (ASEE). It was jointly managed by the California Space Institute and the Lyndon B. JOhnson Space Center, under the direction of the Office of Aeronautics and Space Technology (OAST) at NASA Headquarters. The study participants (listed in the - addendum) included a group of 18 university teachers and researchers (faculty fellows) who were present for the entire 10-week period and a larger group of attendees from universities, Government, and industry who came for a series of four 1-week workshops.

The organization of this report follows that of the summer study. Space Resources consists of a brief overview and four detailed technical volumes: (1) Scenarios; (2) Energy, Power, and Transport; (3) Materials; (4) Social Concerns. Although many of the included papers got their impetus from workshop discussions, most have been written since then, thus allowing the authors to base new applications on established information and tested technology. All these papers have been updated to include the authors' current work.

This overview, drafted by faculty fellow Jim Burke, describes the findings of the summer study, as participants explored the use of space resources in the development of future space activities and defined the necessary research and development that must precede the practical utilization of these resources. Space resources considered included lunar soil, oxygen derived from lunar soil, material retrieved from near-Earth asteroids, abundant sunlight, low gravity, and high vacuum. The study participants analyzed the direct use of these resources, the potential demand for products from them, the techniques for retrieving and processing space resources, the necessary infrastructure, and the economic tradeoffs.

This is certainly not the first report to urge the utilization of space resources in the development of space activities. In fact, Space Resources may be seen as the third of a trilogy of NASA Special Publications reporting such ideas arising from similar studies. It has been preceded by Space Settlements: A Design Study (NASA SP-413) and Space Resources and Space Settlements (NASA SP-428).

And other, contemporaneous reports have responded to the same themes. The National Commission on Space, led by Thomas Paine, in Pioneering the Space Frontier, and the NASA task force led by astronaut Sally Ride, in Leadership and America's Future in Space, also emphasize expansion of the space infrastructure; more detailed exploration of the Moon, Mars, and asteroids; an early start on the development of the technology necessary for using space resources; and systematic development of the skills necessary for long-term human presence in space.

Our report does not represent any Government-authorized view or official NASA policy. NASA's official response to these challenging opportunities must be found in the reports of its Office of Exploration, which was established in 1987. That office's report, released in November 1989, of a 90-day study of possible plans for human exploration of the Moon and Mars is NASA's response to the new initiative proposed by President Bush on July 20, 1989, the 20th anniversary of the Apollo 11 landing on the Moon: "First, for the coming decade, for the 1990s, Space Station Freedom, our critical next step in all our space endeavors. And next, for the new century, back to the Moon, back to the future, and this time, back to stay. And then a journey into tomorrow, a journey to another planet, a manned mission to Mars." This report, Space Resources, offers substantiation for NASA's bid to carry out that new initiative.

Introduction

Future space activities may benefit from the use of natural resources found in space: energy, from the Sun, certain properties of space environments and orbits, and materials of the Moon and near-Earth asteroids. To assess this prospect and to define preparations that could lead to realizing it, a study group convened for 10 weeks in the summer of 1984 at the California Space Institute at the University of California at San Diego. Papers written by this study group were edited and then recycled through most of the contributors for revision and updating to reflect current thinking and new data on these topics. This is a summary report of the group's findings.

The sponsors of the study—NASA and the California Space Institute— charged the study group with the task of defining possible space program objectives and scenarios up to the year 2010 and describing needed technologies and other precursor actions that could lead to the large-scale use of nonterrestrial resources. We examined program goals and options to see where, how, and when space resources could be of most use. We did not evaluate the long-range program options and do not recommend any of them in preference to others. Rather, we concentrated on those near-term actions that would enable intelligent choices among realistic program options in the future. Our central conclusion is that near-Earth resources can indeed foster the growth of human activities in space. Most uses of the resources are within the space program, the net product being capabilities and information useful to our nation both on and off the Earth.

The idea of using the energy, environments, and materials of space to support complex activities in space has been implicit in many proposals and actions both before and during the age of space flight. As illustrated in figure 1, the deep gravity well of the Earth makes it difficult and expensive to haul all material supplies, fuel, and energy sources into space from the surface of the Earth; it is clearly more efficient to make maximum use of space resources. Up to now, however, our ability to employ these resources has been limited by both technology and policy. Studies and laboratory work have failed to bring the subject much beyond the stage of speculations and proposals, primarily because until now there has been no serious intent to establish human communities in space.

With progress in the Soviet program of long-duration manned operations in Earth orbit and with the coming of an American space station initiative, the picture appears to be changing. The present study is one step in a process laying groundwork for the time when living off Earth, making large-scale use of nonterrestrial resources, will be both technologically feasible and socially supported.

Findings of the Summer Study

The 18 faculty fellows who participated in the summer study organized themselves into four groups. The focus of each group corresponded with that of a 1-week workshop held in conjunction with the summer study and attended by 10 to 20 experts in the target field. The first working group generated the three scenarios that formed the basis of the subsequent discussions. The other three groups focused on these areas of inquiry:

Working group 2-Energy, power, and transport

Working group 3-Materials and processing

Working group 4-Human and social concerns

In what follows, our findings are presented in the order of these topics, but they are offered as findings of the summer study as a whole. Integrated in these findings are the views of the faculty fellows and the workshop attendees.

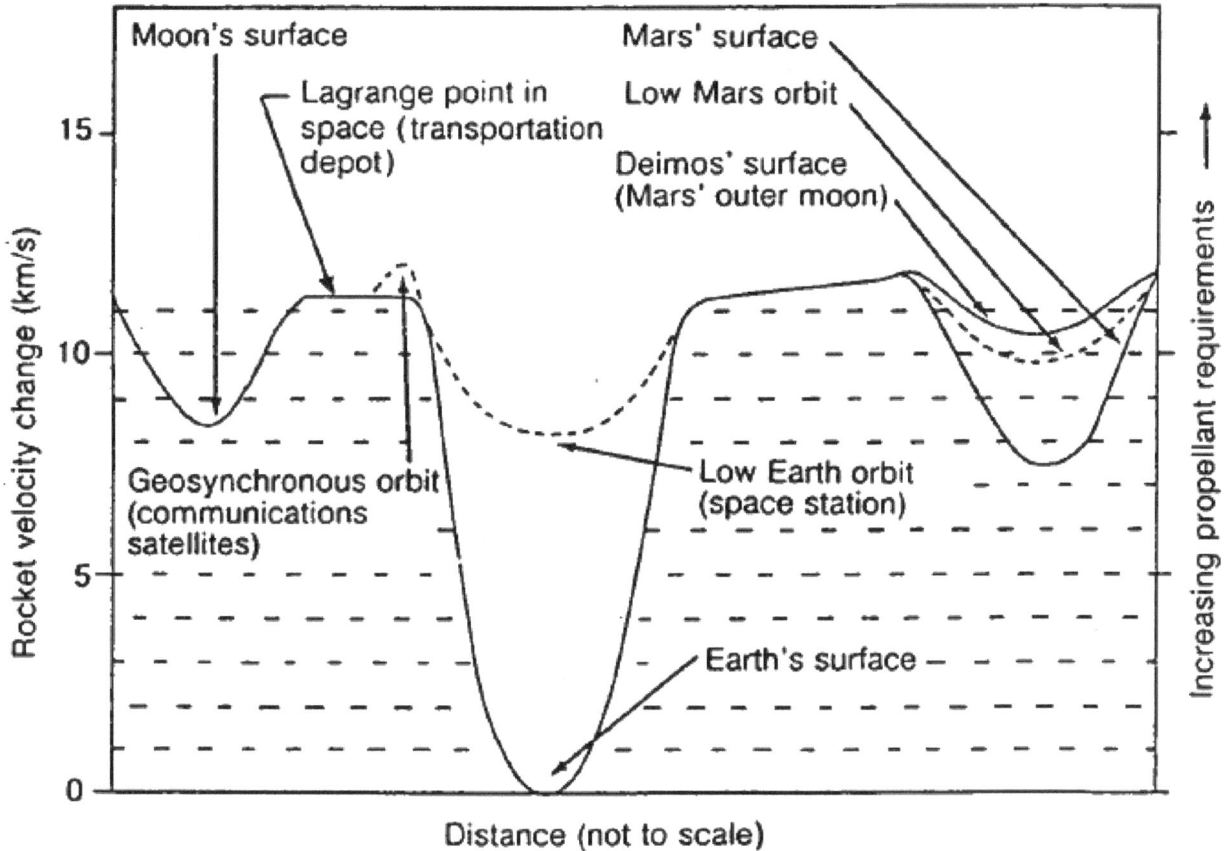

Figure 1

The Gravity Well of the Earth

The Earth sits in a deep gravity well and considerable rocket energy is necessary to lift material from that well and put it into space. The rocket velocity change (ΔV) shown here is an indication of the minimum fuel needed to travel to low Earth orbit and to other places, including the lunar surface and Deimos. Not shown on the diagram but also important is the fact that it takes less ΔV to reach some Earth-crossing asteroids than it does to reach the lunar surface, about 10 percent less for asteroid 1982 08, for example. This diagram is not a potential energy diagram, as the ΔV depends on the path taken as well as the potential energy difference. However, it is a good indication of the relative fuel requirements of transportation from one place to another. The diagram also does not take into account travel times corresponding to minimum AV trajectories. One-way travel times range from less than an hour to low Earth orbit to 3 days for lunar orbit to months to a year or more for Mars and Earth-crossing asteroids.

Future Space Activities

Before we could evaluate the benefits and opportunities associated with the use of space resources, we had to consider what might be going on in space in the future. The target date defined for this study, 2010, is beyond the projection of present American space initiatives but not too far in the future for reasonable technological forecasting. The U.S. space program is now set on a course that can carry it to the end of this century, with increasing capabilities in low Earth orbit (LEO) and geosynchronous Earth orbit (GEO) and modest extensions into deeper space. At the present rate of progress, there would not be much new opportunity to exploit nonterrestrial resources before the year 2000.

A typical plan for space activities is illustrated in figure 2, which shows a sequence of milestones leading to human enterprises in LEO, in GEO, and on the Moon, plus automated probing of some near-Earth asteroids and of Mars. In this plan, most of the space activity before 2010 is concentrated in low Earth orbit, where the basic space station is expanded into a larger complex over a period of 20 years. In geosynchronous Earth orbit, an experimental platform is replaced in 2004 by an outpost to support manned visits leading to a permanently manned station by 2012. Until the year 2010 only unmanned missions are sent to the Moon. In that year, nearly 20 years after the establishment of the space station, a small lunar camp is established to support short visits by people. In this plan, the only American missions to Mars in the next 40 years are two unmanned visits: a sample return mission and a roving surveyor. It is clear that, if the plan in figure 2 is followed, natural resources from the Moon, Mars, or other planetary bodies will not be used until at least 2016.

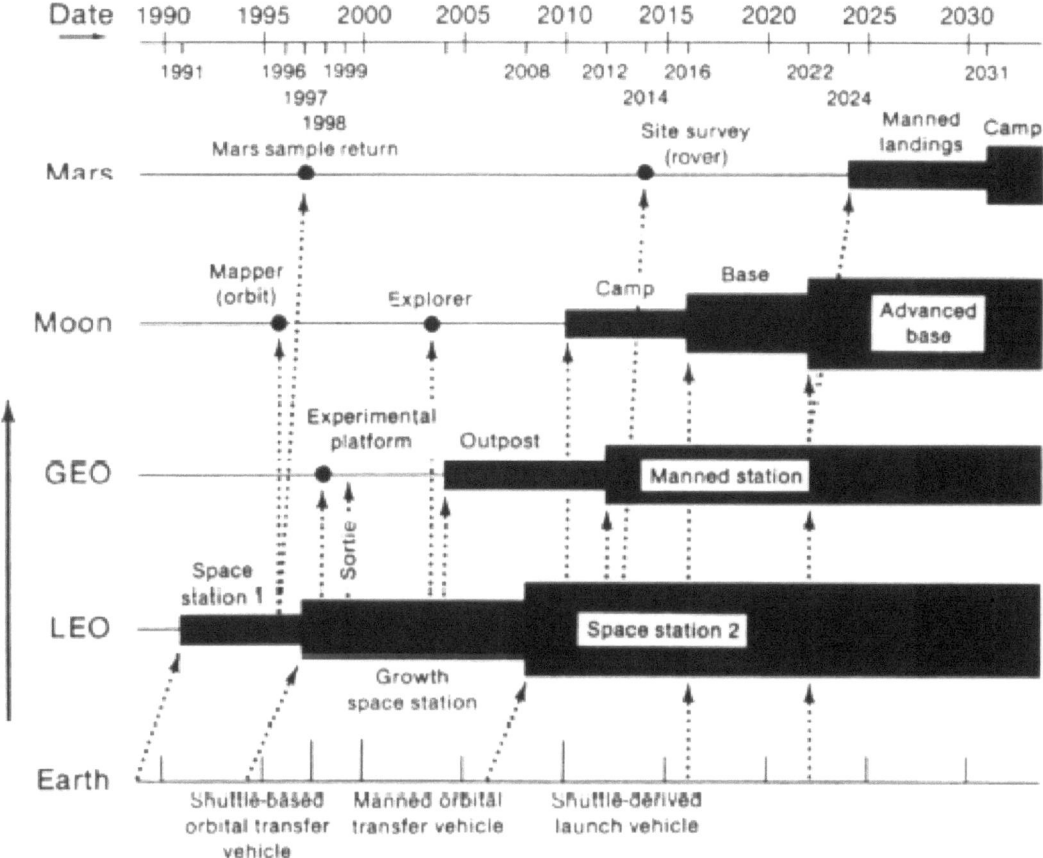

Figure 2

Baseline Scenario

If NASA Continues its business as usual without a major increase in its budget and without using nonterrestrial resources as it expands into space, this is the development that might be expected in the next 25 to 50 years. The plan shows an orderly progression in manned missions from the initial space station in low Earth orbit (LEO) expected in the 1990s, through an outpost and an eventual space station in geosynchronous Earth orbit (GEO) (from 2004 to 2012), to a small lunar base in 2016, and eventually to a Mars landing in 2024. Unmanned precursor missions would include an experiment platform in GEO, lunar mapping and exploration by robot, a Mars sample return, and an automated site survey on Mars. This plan can be used as a baseline scenario against which other, more ambitious plans can be compared.

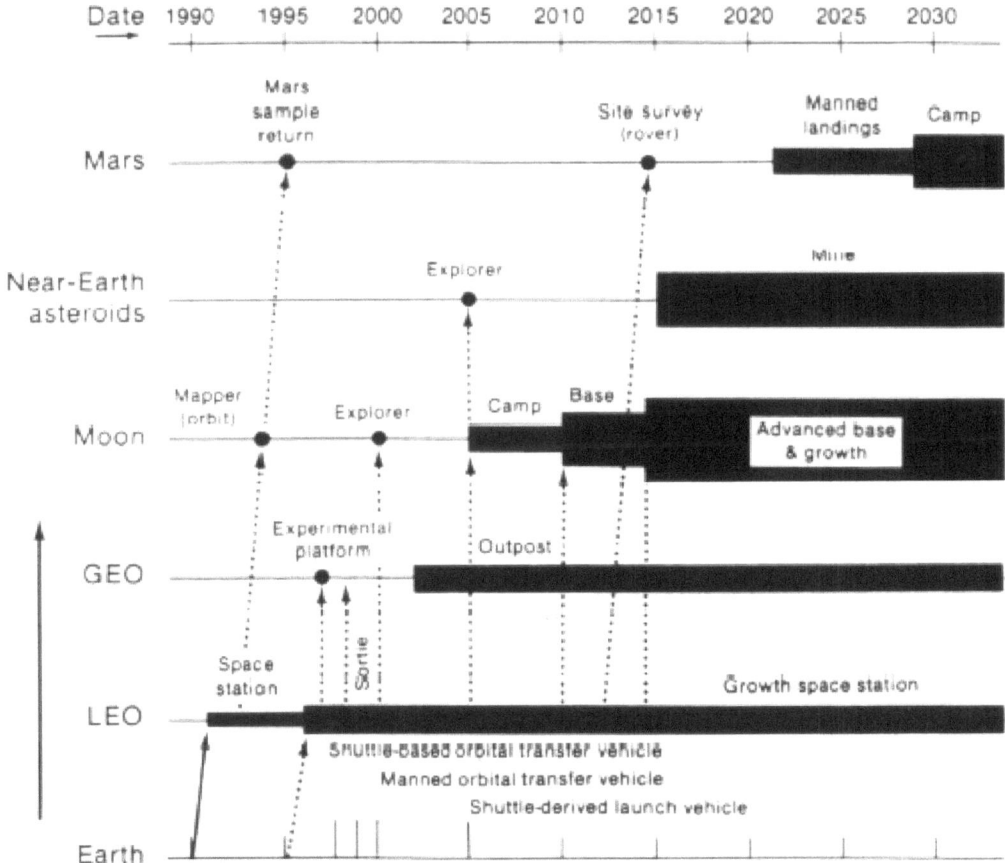

Figure 3

Scenario for Space Resource Utilization

Space resource utilization, a feature lacking in the baseline plan, is emphasized in this plan for space activities in the same 1990-2035 time frame. As in the baseline scenario, a space station in low Earth orbit (LEO) is established in the early 1990s. This space station plays a major role in staging advanced missions to the Moon, beginning about 2005, and in exploring near-Earth asteroids, beginning about the same time. These exploration activities lead to the establishment of a lunar camp and base which produce oxygen and possibly hydrogen for rocket propellant. Automated missions to near-Earth asteroids begin mining these bodies by about 2015, producing water and metals which are returned to geosynchronous Earth orbit (GEO), LEO, lunar orbit, and the lunar surface. Oxygen, hydrogen, and metals derived from the Moon and the near-Earth asteroids are then used to fuel space operations in Earth-Moon space and to build additional space platforms and stations and lunar base facilities. These space resources are also used as fuel and materials for manned Mars missions beginning in 2021. This scenario might initially cost more than the baseline scenario because it takes large investments to put together the facilities necessary to extract and refine space resources. However, this plan has the potential to significantly lower the cost of space operations in the long run by providing from space much of the mass needed for space operations.

If we consider the plan in figure 2 to be our baseline, then figure 3 illustrates an alternative departing from that baseline in the direction of more and earlier use of nonterrestrial resources. In this plan, a growing lunar base has become a major goal after the space station. Lunar and asteroidal resources would be sought and exploited support of this goal rather than for any external purpose. The establishment of a lunar camp is moved up 5 years to 2005 and an advanced lunar base is in place by 2015. In this plan, lunar resources are used to support the construction and operation of this base and lunar-derived oxygen is used to support transportation to and from the base. Asteroidal material from automated mining missions would also contribute to supporting these space operations after 2015.

Figure 4 shows a different departure from the baseline. Here, the objectives are balanced among living off Earth, developing near-Earth resources for a variety of purposes, and further exploring the solar system with an eventual human landing on Mars. In this alternative scenario, a LEO space station, a small manned GEO outpost, and a small manned lunar station are all in operation by 2005, with a manned Mars visit and establishment of a camp by 2010, some 12 to 14 years earlier than in the previous plans. Automated asteroid mining and return starts by 2010. The focus of this program is longer term than that of the program diagramed in figure 3. By building up a balanced infrastructure at various locations, it invests more effort in activities whose benefits occur late in the next century and less in shorter range goals such as maximizing human presence on the Moon.

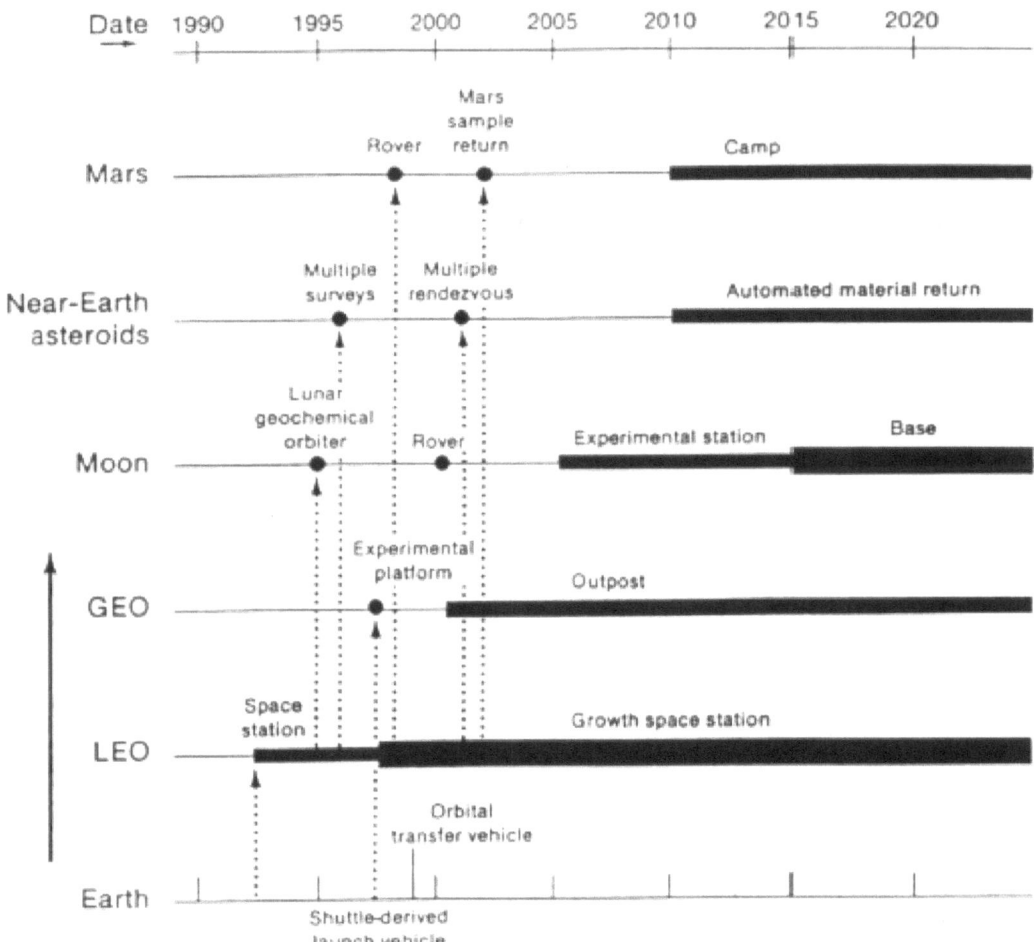

Figure 4

Scenario for Balanced Infrastructure Buildup

In this scenario, each location in space receives attention in a balanced approach and none is emphasized to the exclusion of others. The scenario begins with the establishment of the initial space station about 1992. This is followed by the establishment of a manned outpost in geosynchronous Earth orbit (GEO) in 2001, an experimental station on the Moon in 2006, and a manned Mars camp in 2010. In parallel with these manned activities, many automated missions are flown, including a lunar geochemical orbiter and a lunar rover, multiple surveys of near-Earth asteroids and rendezvous with them, and a martian rover and a Mars sample return. Automated mining of near-Earth asteroids beginning in 2010 is also part of this scenario.

These three scenarios, the baseline and two alternates, have served as a basis for our discussion of the uses of nonterrestrial resources. None is a program recommended by the study group, since that was not our charter. They are merely illustrative examples of programs that, we believe, might materialize over the next two decades as a result of national or international trends in space. The two alternate scenarios assume some acceleration and focusing of American efforts in space, as happened during the Apollo era, while the baseline scenario assumes a straightforward extrapolation of our present program, with only modest budget growth and no particular concentration on the use of nonterrestrial resources.

Heavy Lift Launch Vehicle

An unmanned heavy lift launch vehicle derived from the Space Shuttle to lower the cost of transporting material to Earth orbit would make it feasible to transport to orbit elements of a lunar base or a manned spacecraft destined for Mars. Its first stage would be powered by two solid rocket boosters, shown here after separation. Its second stage would be powered by an engine cluster at the aft end of the fuel tank that forms the central portion of the vehicle. All this pushes the payload module located at the forward end. This payload module can carry payloads up to 30 feet (9.1 meters) in diameter and 60 feet (18.3 meters) in length and up to 5 times as heavy as those carried by the Shuttle orbiter. Artist: Dennis Davidson

Energy, Power, and Transport

We became convinced that a space program large enough to need, and to benefit significantly from, nonterrestrial resources would require a great expansion of energy, power, and transport beyond the capabilities of today. Sunlight is already in use as a primary energy source in space, and nuclear energy has been used on a small scale. Photovoltaic panels, together with chemical or nuclear energy sources brought from Earth, have been sufficient up to now. In the future, more advanced and much larger solar and nuclear energy systems may be built; but, even then, energy supply may limit our rate of progress. For example, a program is under way to develop the SP-100, a space nuclear power plant intended to produce 100 kilowatts of electricity with possible extension to a megawatt. But even a small lunar base would consume several megawatts.

Harnessing sunlight on a large scale and at low cost thus remains a priority research and development goal, as does the creation of high-capacity systems for converting and storing solar and nuclear energy in space. Many studies have described the candidate techniques, including solar furnaces, solar-powered steam engines, solar-pumped lasers, and nuclear thermal power plants.

Although solar energy is ubiquitous and abundant, and compact nuclear energy sources can be brought up from Earth, it is still necessary to have machinery in space for capturing, storing, converting, and using the energy. Perhaps nonterrestrial resources can be used in the creation of some of this machinery. For example, as has often been proposed, lunar silicon could be used for photovoltaics; lunar glass, for mirrors.

A more important energy initiative might be the development of new storage and management concepts, such as the establishment of water, oxygen, and hydrogen caches cryogenically stored in the lunar polar cold traps. Fluidized-bed heat storage, molten metal cooling fountains, and storage by hoisting weights are other examples of energy storage and management benefiting from attributes of the lunar environment; namely, a large supply of raw materials, vacuum, and gravity. Consideration should also be given to the siting of solar and nuclear power plants on the Moon. For example, a solar plant located at one of the lunar poles would be capable of nearly continuous operation, in contrast to a plant at an equatorial location which would be in darkness 14 days Out of 28 (figs. 5 and 6).

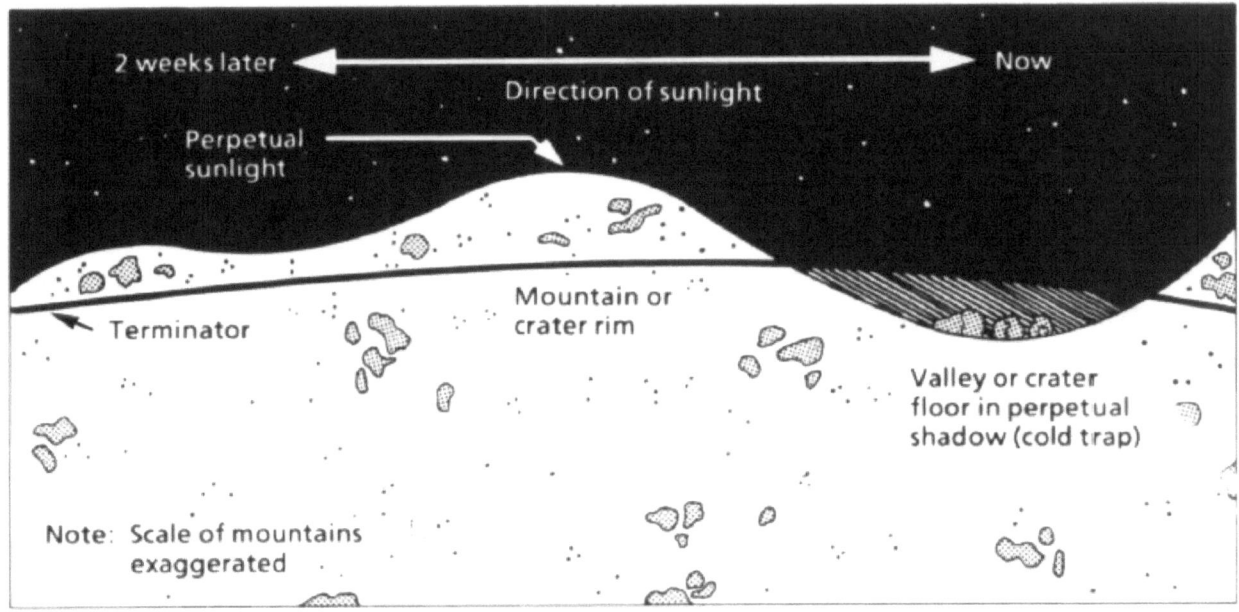

Figure 5

Lunar Polar Illumination

The Moon's diurnal cycle of 14 Earth days of sunlight followed by 14 Earth days of darkness could be a problem for siting a lunar base dependent on solar energy or cryogenic storage. A site that might obviate this problem would be at one of the lunar poles. At a pole, high points, such as mountain tops or crater rims, are almost always in the sunlight and low areas, such as valleys or crater floors, are almost always in the shade. The Sun as seen by an observer at the pole would not set but simply move slowly around the horizon. Thus, a lunar base at a polar location could obtain solar energy continuously by using mirrors or collectors that slowly rotated to follow the Sun. And cryogens, such as liquid oxygen, could be stored in shaded areas with their constant cold temperatures

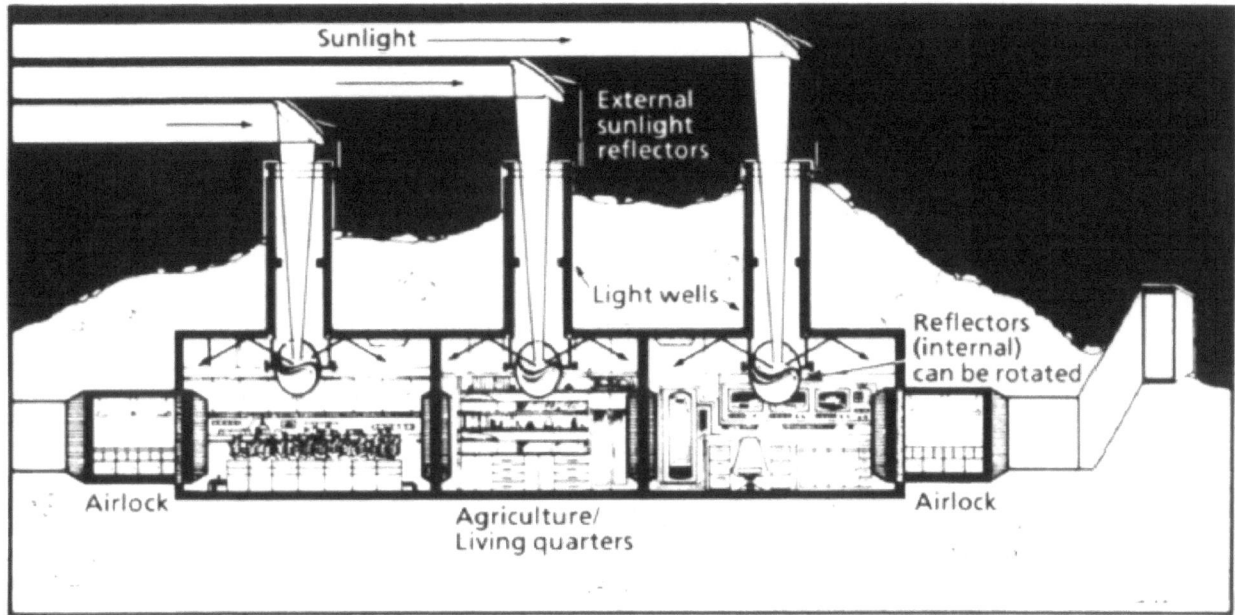

Figure 6

Polar Lunar Base Module

Light could be provided to a lunar base module located at the north (or south) pole by means of rotating mirrors mounted on top of light wells. As the mirrors tracked the Sun, they would reflect sunlight down the light wells into the living quarters, workshops, and agricultural areas. Mirrors at the bottom of the light wells could be used to redirect the sunlight or turn it off.

We found that transport costs would be dominant in any program large enough to make significant use of nonterrestrial resources. We recommend continued pursuit of technologies offering the prospect of large reductions in Earth-to-LEO transport cost. A preliminary economic model of the effect of lunar resource utilization on the cost of transportation in space was developed by the study. This model, developed in more detail, shows that delivery of lunar-derived oxygen to LEO for use as propellant in space operations would be significantly cheaper than delivery of the same payload by the Space Shuttle, assuming a demand for about 300 metric tons of oxygen delivered to LEO. If Earth-to-LEO costs could be reduced using unmanned cargo rockets - Shuttle-derived launch vehicles or heavy lift launch vehicles, the cost of lunar-derived oxygen would also be reduced. At this demand level, if Earth-to-LEO costs were lowered to about 2/3 their present value, it would be cheaper to bring all oxygen up from Earth. But, if demand for liquid oxygen as propellant in LEO were to grow by a factor of 2 or more, then lunar-derived oxygen would be competitive with Earth-derived oxygen using any currently contemplated launch vehicle. This model points Out that considerable reduction in unit cost for lunarderived oxygen delivered to LEO can be achieved as the volume and scale of operations increase. The model assumes that all hydrogen is transported from Earth. If lunarderived hydrogen were available, the cost of providing lunar-derived oxygen would be considerably reduced at all production rates.

While enhancing Earth-to-orbit capacity, we should be preparing to expand our range. For example, an orbital transfer vehicle (ON) is needed for traffic to and from GEO. Extending the space-based ON concept to meet the needs of a lunar base transport system should be considered from the outset of OTV development. The development of an efficient ON capable of LEO-to-Moon transportation was identified by the economic model just summarized as the single most important factor in the cost of supplying lunar-derived oxygen to space operations.

Also, we support the findings of other studies, such as NASA's 1979 report Space Resources and Space Settlements, to the effect that it may be practical and desirable to transport lunar material using means other than the OW-derived vehicles that will be carrying humans to and from the Moon. The lunar environment encourages consideration and development of electromagnetic launchers and other unconventional transport devices.

We recognize a need for transport of both equipment and personnel from place to place on the lunar surface and probably also a need for at least short-range transport of raw and processed lunar materials. Much of this transport would logically be provided by teleoperated vehicles. Teleoperated systems, robotics, and automation developed for the space station may have direct application in lunar operations. Such systems would be absolutely required by any program to mine and utilize material from near-Earth asteroids.

Finally, we recommend that alternative advanced propulsion technologies be developed to permit comparison and selection of systems for transport beyond Earth orbit. Examples include solar thermal propulsion, solar electric ion thrusters, nuclear electric propulsion, laser-powered systems, and light-pressure sailing.

Locations, Environments, and Orbits

Another natural resource is afforded by orbits and places in the solar system. The geosynchronous orbit, used for communications and observation, is a resource that has led to the largest commercial development in space and offers an even greater payoff in the future. The combined gravity fields of the Sun and planets offer a resource that has already been used for modifying and controlling spacecraft trajectories through swing-by maneuvers. Aeromaneuvering in planetary atmospheres and momentum exchanges using tethers offer additional means of trajectory control.

In future space activities, unique space environments may become important resources. Examples include the Moon's far side, which is shielded from the radio noise of Earth and would thus be an excellent location for a deep-space radio telescope. Lunar orbit or the gravitationally stable Lagrangian points in the Earth-Moon-Sun system may be good locations for space platforms. As has already been pointed out, the lunar poles have the potential of providing constant sunlight to power a lunar base. And aerobraking in the Earth's upper atmosphere may make it possible to bring both lunar and asteroidal material into low Earth orbit for use in space activities.

We found that any future program intending to make major use of nonterrestrial resources, especially the materials of the Moon, must include a substantial human presence beyond Earth orbit. This finding leads to the conclusion that some form of extended human living in deep space, such as a lunar base, is a necessity (fig. 7).

The space station is the obvious place to conduct the proving experiments that will enable confident progress toward productive lunar living, including use of local resources. While this summer study group did not attempt to lay out an entire plan of events leading up to establishment of a lunar base, we recognized some of the steps that are logical and likely to be considered essential. One of these is a suite of experiments, in the space station, demonstrating the soundness of methods and processes to be used at the lunar base.

Since a number of these methods and processes are gravity-dependent, it is necessary to demonstrate them at simulated lunar gravity, 1/6 g, and this cannot be done on Earth. We therefore recommend that space station facilities include a 1/6 g centrifuge in which lunar base experiments and confirmation tests can be carried out.

Lunar Orbit Space Station

Proximity to lunar-derived propellant and materials would make a space station in orbit around the Moon an important transportation node. It could serve as a turnaround station for lunar landing vehicles which could ferry up liquid oxygen and other materials from the lunar surface. An orbital transfer vehicle could then take the containers of liquid oxygen (and possibly lunar hydrogen) to geosynchronous or low Earth orbit for use in many kinds of space activities. A lunar orbit space station might also serve as a staging point for major expeditions to other parts of the solar system, including Mars. Artist: Michael Carroll

Figure 7

Advanced Lunar Base

In this artist's conception of a lunar base, a processing plant in the foreground is producing oxygen and fused glass bricks from lunar rocks and soil. The rocks and soil are fed into the system on the left side from a robot-controlled cart. Solar energy concentrated by the mirror system is used to heat, fuse, and partially vaporize the lunar material. The oxygen-depleted fused soil is cast into bricks, which are used as building blocks, paving stones, and radiation shielding. Oxygen extracted from the vapor is piped to an underground cryogenic plant, where it is liquefied and put into the round containers shown under the shed. The rocket in the background will carry these containers into space, where the oxygen will be used as rocket propellant. The lunar base living quarters are underground in the area on the left. The solar lighting system and the airlock entry are visible. As this lunar base expands, additional useful products such as iron, aluminum, and silicon could be extracted from the lunar rocks and soil.

Polar Solar Power System

At a base near a lunar pole, a solar reflector (the large tower in the background) directs sunlight to a heat collector, where it heats a working fluid which is used to run a turbine generator buried beneath the surface. At such a location the solar power tower can track the Sun simply by rotating around its vertical axis. Power is thus provided continuously without the 2-week nighttime period which is characteristic of nonpolar locations. The triangle in the background is the mining pit. In the foreground, two scientists collect rock samples for analysis at the base. Artist: Maralyn Vicary

Materials and Processing

Any material that is already in space has enormous potential value relative to the same material that needs to be brought up from Earth, simply because of the high cost of lifting anything out of the Earth's deep gravity well. On an energy basis, it is more than 10 times as easy to bring an object into low Earth orbit from the surface of the Moon as from the surface of the Earth (see figure 1). Residual propellants and tanks or other hardware left in orbit can constitute a resource simply because of the energy previously invested in them.

These facts of nature underlie many proposals for the use of nonterrestrial materials. For example, as discussed in the transportation section, there could be a payoff if lunar oxygen, abundant in the silicates and oxides of the Moon and extractable by processes conceptually known, were to be used in large quantities for propulsion and life support in space operations. A sketch of a concept for extracting oxygen from lunar materials is shown in figure 8. Byproducts of this process might include useful metals.

The materials of near-Earth asteroids may complement the materials of the Moon. On the basis of evidence gained to date, the Moon is lacking in water and carbon compounds—important substances that are abundant in certain classes of meteorites and thus may be found among the small asteroids that orbit the Sun near us. Water from asteroids could provide hydrogen for use as rocket fuel in space operations. On an energy basis, many of the near-Earth asteroids are even easier to reach than the Moon. And there are more energy advantages in a payload return from an asteroid because of their very low gravity. This same low gravity may require novel techniques for mining asteroids (figs. 9 and 10). Low-energy transit times to asteroids are relatively long (months or years, in comparison to days for the Moon), so the voyages to obtain these asteroid materials will probably be automated rather than manned.

Figure 8

Solar Furnace Processing of Lunar Soil To Produce Oxygen

A device like this could utilize solar energy to extract oxygen from lunar soil. Lunar soil is fed into the reactor through the pipe on the left. Concentrated solar rays heat the soil in the furnace. Hydrogen gas piped into the device reacts with ilmenite in the soil, extracting oxygen from this mineral and forming water vapor. Ilmenite, an iron-titanium oxide, is common in lunar mare basalts. When this mineral is exposed to hydrogen at elevated temperatures (around 900° C), the following reaction takes place:

$$FeTiO_3 + H_2 = Fe \text{ (metal) } TiO_2 + H_2O$$

In the device illustrated, the water vapor is removed by the unit on the right and electrolyzed to yield oxygen gas and hydrogen gas. The hydrogen gas is cycled back into the reactor. The oxygen gas is cooled and turned into liquid oxygen. Metallic iron is a useful byproduct of this reaction. The production of liquid oxygen for life support and propellant use, both on the Moon and in Earth-Moon space, is such an important economic factor that it could enable a lunar base to pay for itself.

Our first finding with regard to materials and processing is obvious but still needs to be stated explicitly because it is so important.

The United States will have no access to nonterrestrial materials unless there is a substantial change in the national space program. Because of recent budget limits and a concentration on applications in LEO and GEO, we have no capability to send humans to the Moon. The option of an entirely automated lunar materials recovery operation, while it might be technically feasible, appears to us unlikely to gain approval. With regard to asteroidal resources, automated return to Earth orbit is mandated by trip times, but the processing would still require human supervision. These findings have two consequences: first, that the utilization of nonterrestrial materials awaits the creation of some new system for high-capacity transport beyond LEO; and, second, that large-scale utilization awaits the creation of a lunar base or an asteroid mining and recovery scenario.

Our other findings regarding lunar, asteroidal, and martian materials presume that the nation has found some way to get over the hurdles just described. With the required transport and habitat infrastructure in place, the question reduces to one of considering possible ways to process and use the materials.

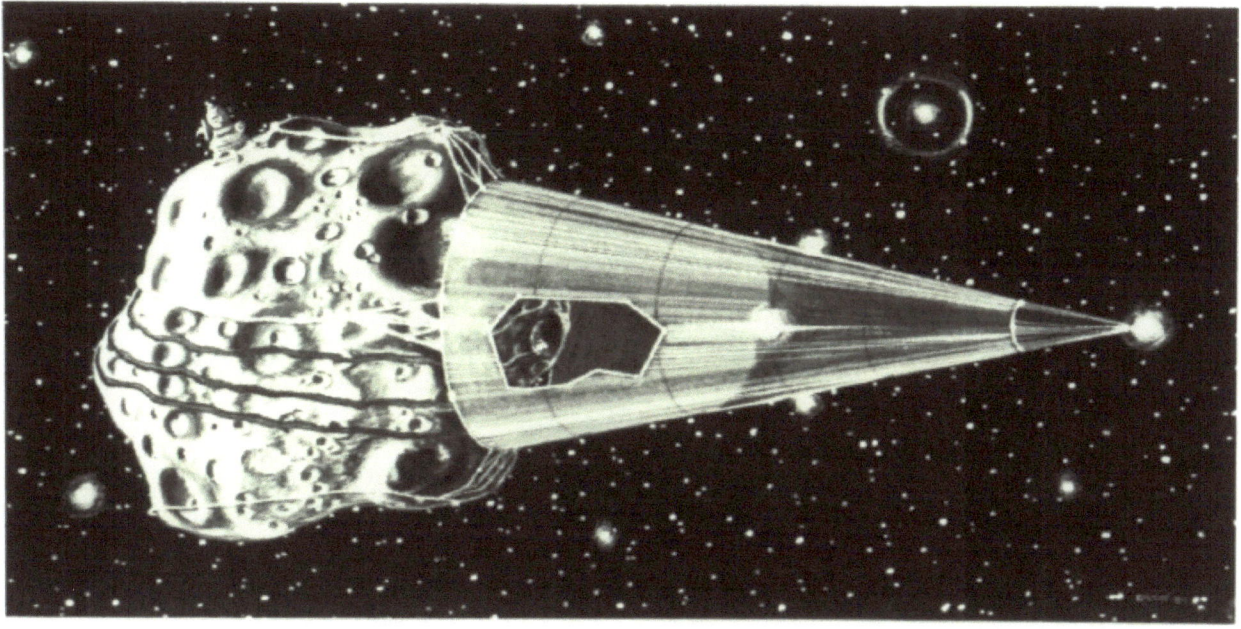

Figure 9

Tethered Asteroid

In this drawing, a small asteroid is being mined for raw materials from which water and metals can be extracted. A landing craft is shown on the surface near the top of the asteroid. Robotic devices from this craft have attached a large cone-shaped shroud with tethers which go completely around the asteroid. A small mining vehicle (see next figure), also held to the surface with tethers, uses paddle wheels to throw loose asteroid regolith up from the surface. The regolith is caught by the shroud. When full, the shroud can be propelled by attached engines or towed by another vehicle to a processing plant in Earth orbit.

Figure 10

Asteroid Mining Vehicle

Because of asteroids' extremely low gravity, normal mining methods (scoops, etc.) may not be practical and unusual methods may be required. This innovative asteroid mining vehicle is designed to be used with the shroud shown in the previous figure. As the robotic vehicle moves across the asteroid, it is held to the surface by tethers. The vehicle has rotating paddle wheels that dig into the regolith and throw loose material Out from the asteroid to be caught by the shroud. Other techniques might also be tried, such as using a tethered or anchored rig to drill or melt big holes. Tradeoff studies must be made to determine whether it is more efficient to process the raw material into useful products on or near the asteroid or to bring back only raw material to a processing plant near Earth. Because of very long transportation times (up to several years for round trips), it is probably not practical to have asteroid mining missions run by human crews. Automated, robotic, and teleoperated missions seem more practical. However, the complexity of such a mission is likely to be high.

Lunar oxygen, raw lunar soil, lunar concretes," lunar and asteroidal metals, and asteroidal carbonaceous and volatile substances may all play a part in the space economy of the future.

Because oxygen typically constitutes more than three-quarter of the total mass launched from Earth, an economical lunar oxygen source would greatly reduce Earth-based lift demands. Since transport to LEO accounts for a major portion of the total program cost, use of nonterrestrial propellants may permit faster growth of any program at a given budget level.

Another potential use of nonterrestrial resources is in construction, ranging from the simple use of raw lunar soil as shielding to the creation of refined industrial products for building large structures in space.

At the outset, we believe that lunar material will be used rather crudely; for example, by piling it on top of habitat structures brought from Earth. Even that conceptually simple use implies a significant dirt-moving capacity on the Moon. In any event, the use of local material for radiation and thermal shielding is probably essential because of the prohibitive cost of bringing up an equivalent mass from Earth.

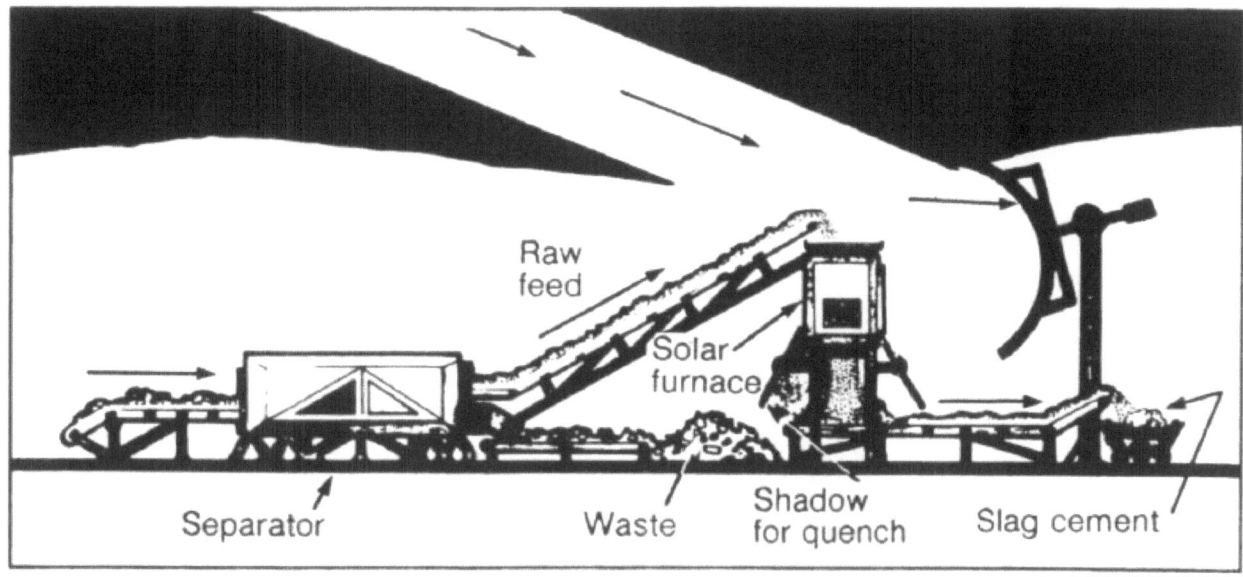

Figure 11

Slag Cement Production Facility

It seems possible to make a usable cement on the lunar surface by relatively simple means. Feedstock separated from lunar soil would be melted in a solar furnace and then quenched in shadow to form a reactive glassy product. When this product is mixed with water and aggregate and allowed to react and dry, it should make a coherent concrete suitable for many structures at a lunar base.

Going beyond just raw soil, it is reasonable to ask whether or not a structural material equivalent to concrete could be created on the Moon. Studies by experts in the cement industry suggest that lunar concrete is a possibility, especially if large amounts of energy and some water are available (figs. 11 and 12). Even without water, it may be possible to process lunar soil into forms having compressive and shear strength, hence usable in structures. Examples include sintered soil bricks, cast glass products, and fiberglass.

Metals are also available on the Moon and asteroids. Metallic iron-nickel is a major component of most meteorites and probably most asteroids. Meteoritic iron, extracted magnetically from lunar soils, can be melted and used directly. Ultrapure iron, which could be produced in the Moon's vacuum and which would not rust even in the moist oxygenated air of a lunar habitat, may prove to be a valuable structural material. Other metals, including titanium and aluminum, are abundant on the Moon but are bound in oxides and silicates so that their extraction is more difficult.

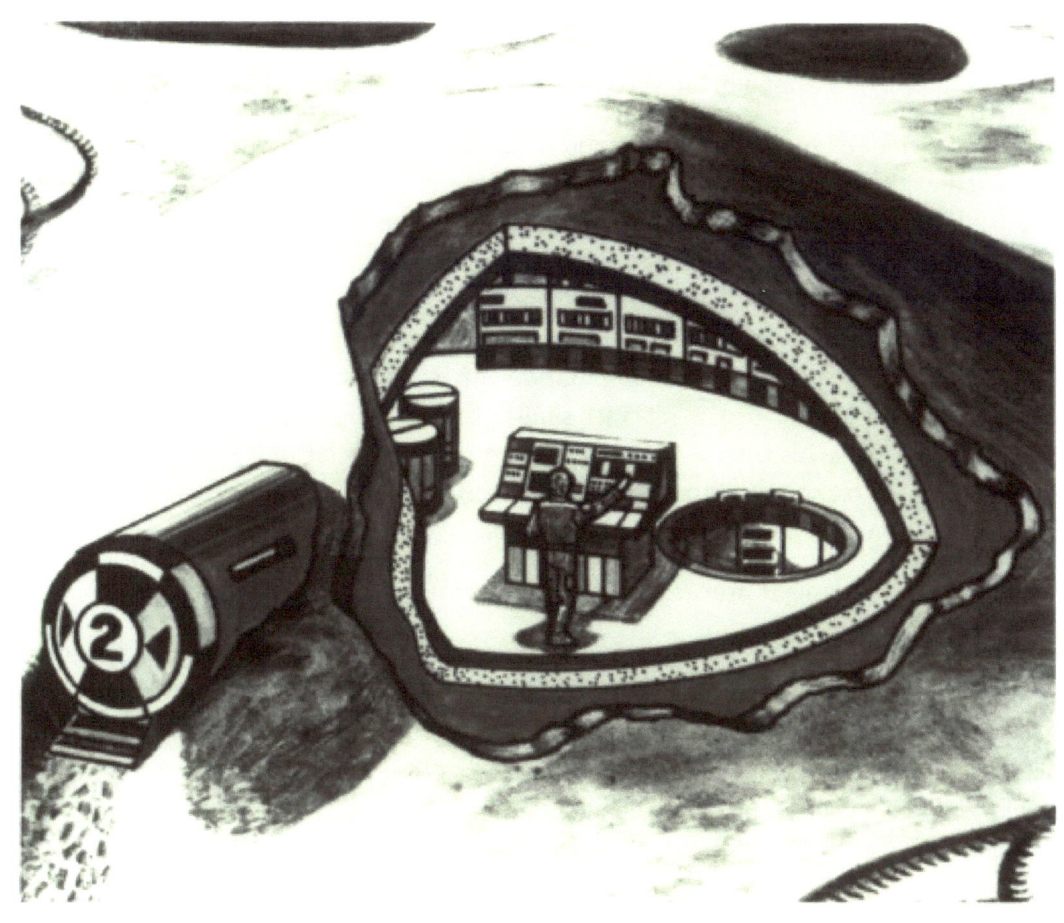

Figure 12

Lunar Base Control Room Made of Lunar Concrete

One use for concrete made primarily from lunar resources is seen in this cutaway sketch of a control room at a lunar base. Together with a blanket of lunar regolith, concrete would provide excellent shielding from the cosmic rays and solar flares that would be a serious hazard at a lunar base designed for longterm habitation. Such shielding could also be used to protect facilities in LEO or GEO.

In an early lunar base, the air, water, and food to support human life will have to be supplied from Earth. As experience is gained, both in a LEO space station and on the Moon, recycling will become more practical, allowing partial closure of the life support system and greatly reducing resupply needs. At some point, local raw materials can be introduced into the cycle. This may be one of the first uses of lunar oxygen and of hydrogen implanted in lunar soil by the solar wind. Then, on a larger scale, lunar materials may be used as a substrate and nutrient source for agriculture. Asteroids can supply substances, such as carbon compounds and water, in which the Moon is deficient. Asteroidal water may be particularly valuable, if no ice is discovered on the Moon and if the hydrogen trapped in lunar soil proves to be impractical to utilize.

A more complete understanding of lunar and asteroidal resources will require additional exploration. Such exploration can be done without making any decision to commit to utilization of nonterrestrial resources and will provide important new data which will help in making such decisions. We therefore recommend that NASA's Office of Space Science and Applications (OSSA) and Office of Space Flight (OSF) jointly sponsor and conduct the study, analysis, and advocacy of two automated flight missions: a lunar polar geochemical orbiter and a near-Earth asteroid rendezvous, each having a combination of scientific and resource-exploration objectives. Both missions could use spacecraft similar to the Mars Observer now planned for launch in the early 1990s. Also, to evaluate the resource potential of Mars and its moons, Phobos and Deimos, we recommend that the Mars Observer data analysis be planned to include resource aspects, such as the potential for in situ propellant production.

Lunar resource exploration might proceed in one of three ways:

- A straight return by a human crew to a site on the Moon where features have been explored and sampled (such as that of Apollo 15, 16, or 17), with the intent of starting base buildup and resource utilization at that site; or
- Establishment of a prebase camp at some other site on the Moon, with the intent that humans would evaluate the local resources; or
- Conduct of an automated, mobile lunar surface exploration mission as a precursor to base siting.

Since strategy can be a function of the discoveries of a remote-sensing mission, we offer no recommendation regarding the choice among these options. We do, however, recommend that early lunar base plans allow for the possibility that any of the options might prove best.

Once serious planning for the use of a particular body of lunar material begins, it will be necessary to determine the extent of the potential mine in three dimensions.

New instruments for probing to modest depths beneath the lunar surface may be required. We therefore recommend that limited-depth mapping be included among the objectives of any lunar surface exploration mission.

Asteroidal exploration might proceed by sending an automated lander and sample return mission to the most favorable near-Earth asteroid. The asteroid rendezvous would have to be preceded by an Earth-based search for the right asteroid. The search for near-Earth asteroids, and their characterization by remote sensing using ground based telescopes, is a good example of a scientific activity with strong implications for the use of nonterrestrial resources. This work is now going on with a mixture of private and public support and could readily be accelerated at low cost.

Laboratory research, on a relatively small scale, using lunar simulants could yield fundamental knowledge important in choosing which technology to develop for the extraction of lunar oxygen, hydrogen, and metals. At similar levels, useful research could be done using meteorites to assess the technology needed to process asteroidal materials for water, carbon, nitrogen, and other volatiles. We recommend that NASA encourage such materials research.

It is a finding of the present study that the processing of nonterrestrial materials, though conceptually understood, has yet to be reduced to practice despite numerous past studies, recommendations, and even some laboratory work. In view of the long lead times characteristic of projects bringing new raw materials sources into production, we believe that more active preparations will soon be needed.

Though laboratory research in this area, as outlined above, is necessary, there are some processes that are ready for technology development and competitive evaluation at pilot-plant scale both on Earth and in space. A logical next step would be processing demonstrations at reduced gravity in the space station and ultimately on the Moon. An example of the needed technology would be a solar furnace designed to extract oxygen and structural materials from lunar soil on the surface of the Moon (see figure 8).

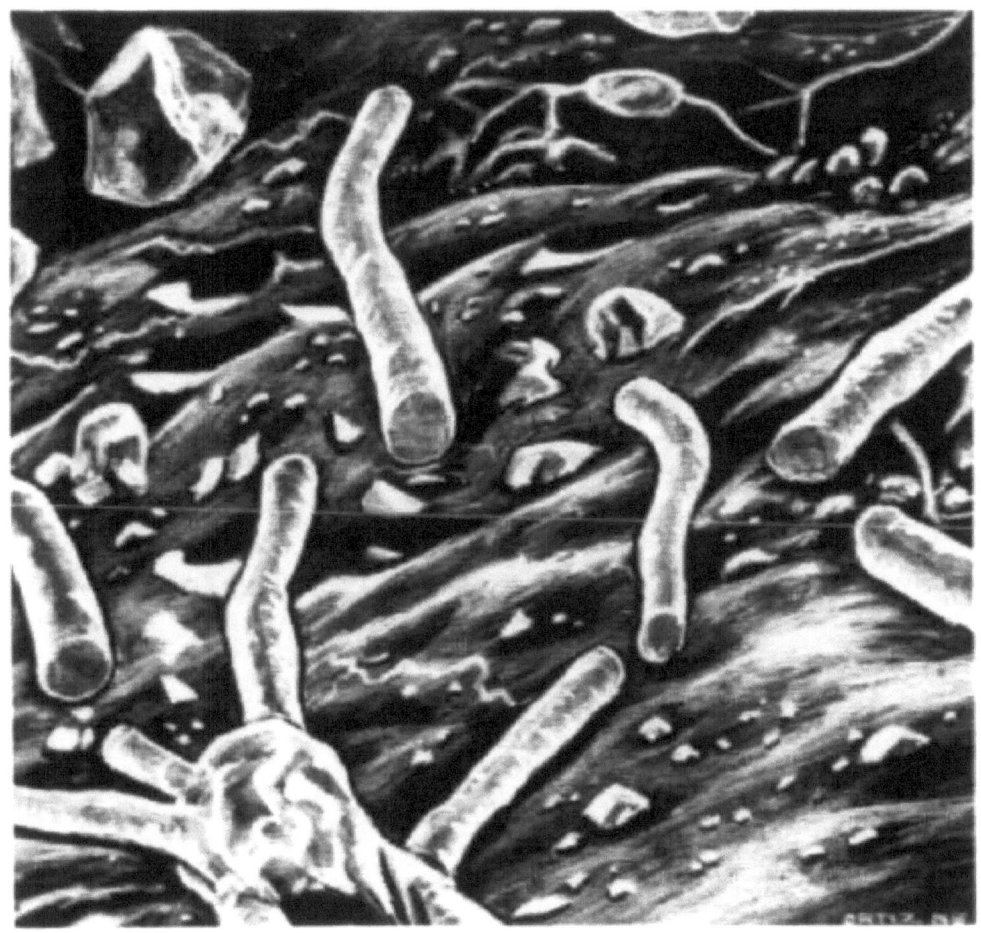

Figure 13

Bacterial Processing of Metal Ores

Although most concepts of processing lunar and asteroidal resources involve chemical reactors and techniques based on industrial chemical processing, it is also possible that innovative techniques might be used to process such resources. Shown here are rod-shaped bacteria leaching metals from orebearing rocks through their metabolic activities. Bacteria are already used on Earth to help process copper ores. Advances in genetic engineering may make it possible to design bacteria specifically tailored to aid in the recovery of iron, titanium, magnesium, and aluminum from lunar soil or asteroidal regolith. Biological processing promotes the efficacy of the chemical processes in ore beneficiation (a synergistic effect).

So much remains unknown about the behavior of the living systems (humans, microorganisms, plants, and animals) that will occupy the space habitats of the future that this is a research field with a very likely payoff. As in the case of inorganic materials, some aspects of this problem have already come past the research stage and are ready for technology development and evaluation. We recommend that NASA's Office of Aeronautics and Space Technology (OAST) support biotechnology work in two areas: (1) plant life support and intensive agriculture under simulated lunar conditions, leading to experimental demonstrations on a 1/6 g centrifuge in the space station, and (2) biological processing of natural raw materials, lunar and meteoritic, to concentrate useful substances (fig. 13). Some such techniques are already in use on a large scale in the mining industry on Earth.

Products derived from the processing of space resources will be used mainly or entirely in the space program itself, at least up to our reference date of 2010. Plans and methods should be developed with this in mind. We do not find any early application of nonterrestrial materials or products made from them on the surface of the Earth. Rather, these materials can accelerate progress at any given annual budget level and thus increase the space program's output of new information, which continues to be its main product.

We found that, while Mars and its moons (fig. 14) almost surely provide a large resource and thus offer the best prospects for sustained human habitation, the most likely use of martian resources would be local; that is, in support of martian exploration and settlement rather than for purposes elsewhere.

Figure 14

Phobos

One of the two moons of Mars, Phobos is slightly ellipsoidal, measuring about 25 km by 21 km. The surface of this moon is heavily cratered and grooved The grooves may be the surface expressions of giant fractures in Phobos caused by an impact that nearly tore it apart and formed the large crater Stickney. The reflectivity of Phobos is similar to that of a type of asteroids that are thought by some to be made of carbonaceous chondrite material. If Phobos is indeed made of this material, it is likely to be rich in water and other volatiles. The loose, fine-grained regolith on the surface appears to be several hundred meters thick in places. This regolith might be relatively easy to mine and process for propellants such as oxygen and hydrogen. Metals might also be extracted from it Similar techniques might be used to mine near-Earth asteroids for propel/ants or metals. These small bodies are of interest because less rocket energy is required to reach and return from some of them than is required to travel to the Moon and back.

Human and Social Concerns

Of all the natural resources in space, the most important in the long run will be the humans living there. Once working settlements (as distinct from expeditions) are established off the Earth, there will be opportunities for qualitative changes in human culture—in the space settlements and in the supporting communities on Earth.

Technologies must be developed to help people get into space, explore it, and live in it. And the use of nonterrestrial resources will affect the development of these technical changes. Agriculture is a clear example: until food production is achieved off Earth, human settlements will remain only outposts utterly dependent on resupply. Thus, the conversion of nonterrestrial materials into substrates for plant growth and the development of food plants usable off Earth will be primary needs.

More important, these technical changes can lead to cultural changes that will improve the quality of life for all space inhabitants. The United States and the Union of Soviet Socialist Republics are now taking the first steps toward permanent habitation of space. If this trend continues, it can divert some of both nations' high technology resources into efforts that are no threat to the people of Earth, and it can lead to the development of human courage, self-reliance, disciplined thinking, and new skills, on a scale otherwise known only in war. These human attributes can be the ultimate product of a program using what nature has provided off the Earth.

It appears likely that future projects will have large capital demands at the outset, large-scale management problems, and high risk both to capital and to national prestige. However, they may offer big economic rewards and many possible nonfinancial rewards, including extension of the human presence in space, development of new culture, and ultimately perhaps even favorable changes in the human species. We can expect scientific advances leading to greater technological excellence; the transfer of new ideas, knowledge, and technology to the Earth; new entrepreneurial horizons; the discovery of unpredicted resources; as well as unprecedented explorations and novel human experiences; opportunities for international cooperation; and the enhancement of American prestige and leadership.

Crisis at the Lunar Base

A projectile has penetrated the roof of one of the lunar base modules and the air is rapidly escaping. Three workers are trying to get into an emergency safe room, which can be independently pressurized with air. Two people in an adjoining room prepare to rescue their fellow workers. The remains of the projectile can be seen on the floor of the room. This projectile is probably a lunar rock ejected by a meteorite impact several kilometers from the base. A primary meteorite would likely be completely melted or vaporized by its high-velocity impact into the module, but a secondary lunar projectile would likely be going slowly enough that some of it would remain intact after penetrating the roof. Detailed safety studies are necessary to determine whether such a meteorite strike (Or hardware failure or human error) is likely to create a loss-of-pressure emergency that must be allowed for in lunar base design. The presence of small safety chambers like this one would perhaps be useful as reassurance to lunar base occupants even though they were never actually used. Artist: Pamela Lee

Our central finding in this area is that, as the space program advances to a state where nonterrestrial resources can be used, its human aspects will become more and more important. The use of Earth's resources, both on land and on and under the sea, provides a clear example and suggests that many of these human problems—legal, political, environmental —will prove difficult and thus will demand early attention. Even if these problems are solved, there will remain substantial human problems within the program. Not only life support but also opportunities for the creative exercise of human talent off Earth must be provided if we are to reach a state where the use of nonterrestrial resources begins to yield a net gain to civilization.

We recommend that NASA encourage (and where possible sponsor) laboratory-scale research on the fundamentals of living systems, with the aim of improving the basis for choices in larger scale efforts such as the controlled ecological life support systems (CELSS) program and space station life support development. Habitat concepts should be studied, including resource substitutions and self-sufficiency to reduce resupply demand. In the specific context of this study, we recommend that this work consider the prospect of using lunar resources (both materials and environments) and asteroidal raw materials to support living systems on the Moon. Design studies should be made of generic human-machine systems adaptable to multiple locations off Earth and able to use local resources to the greatest extent feasible.

Robotics, automation, information, and communications—subjects already important in OAST's programs—will clearly be technologies both driven by and enabling the use of space resources. We recommend that OAST examine, and modify as appropriate, the ongoing NASA robotics, automation, information, and communications technology program in respect to those aspects affecting, or affected by, the use of nonterrestrial resources. An example could be the technology of lunar-surface-based teteoperators for mining and material transport.

Once people are established in low Earth orbit, a whole new field of engineering will begin to grow: operations centered off Earth. Experience with manned and automated operations controlled from centers on Earth shows that the operations discipline is a demanding and expensive one, often rivaling the hardware and other cost elements of a flight project and typically involving hundreds of skilled people acting in a carefully orchestrated manner. Technology can do much to reduce operations costs, but, even with Earth basing, realizing this potential has proved to be difficult. Advance simulations of space-based operations will probably pay dividends.

We recommend that OAST examine cost sources in present day operations and investigate ways to reduce costs of space-based operations including mission control, maintenance and repair, refueling, and logistics and storage, using nonterrestrial resources where appropriate.

Basic research in support of the management of space-based operations should be carried out in biosocial systems, including general living systems research (see figure 15) and consideration of cognitive psychology, management science, the human migration process, and modes of human cooperation in space.

Funding for life support and general living systems research could be found at NASA's Office of Aeronautics and Space Technology (OAST), the National Science Foundation, the National Institutes of Health, the Environmental Protection Agency, and the Office of Naval Research. Ergonomics or human factors research could be funded by NASA OAST, the National Science Foundation, the National Institutes of Health, the Navy, Air Force, or Army, or the Department of Transportation. Space law and policy studies could be funded by the Law and Social Sciences Division of the National Science Foundation or the Office of Commercial Space Transportation of the Department of Transportation. Our point is that research on extended human presence in space requires expanded public and private support. Nationally, for example, other Government agencies beyond NASA should be investing in space R&D, as well as corporations and foundations outside the aerospace industry. International investment in such research is also in order.

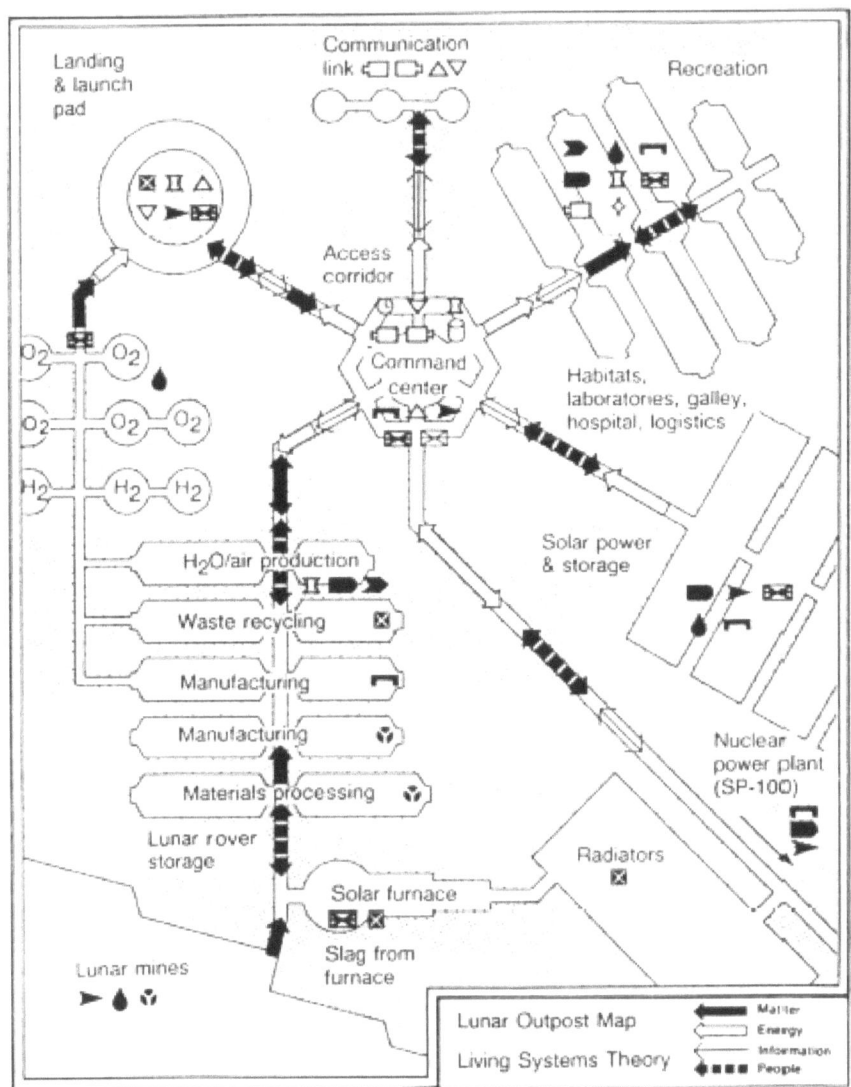

Figure 15

Lunar Outpost Map

General living systems theory and analysis constitutes a rational way to begin to understand the human factors which should guide all planning of space missions. This theory is a conceptual integration of biological and social approaches to the study of living systems. Living systems are open systems that input, process, and output matter and energy, as well as information which guides and controls all their parts. In human organizations, in addition to matter and energy flows, there are flows of personnel, which involve both matter and energy but also include information stored in each person's memory. There are two types of information flows in organizations: human and machine communications and money or money equivalents. Twenty subsystem processes dealing with these flows are essential for survival of systems at all levels.

General living systems theory has been applied in studying such organizations as corporations, military units, hospitals, and universities. it can similarly be applied in studying human settlements in space. The general procedure for analyzing such systems is to map them in two- or three-dimensional space. This map of a lunar outpost indicates its subsystems and the major flows within it. When these flows have been identified, gauges or sensors can be placed at various locations throughout the system to measure the rate of flow and provide information to each of the inhabitants and to others about the processes of the total system, so that its management can be improved (management information system) and its activities made not only more cost-effective but also more satisfying to the humans who live in it.

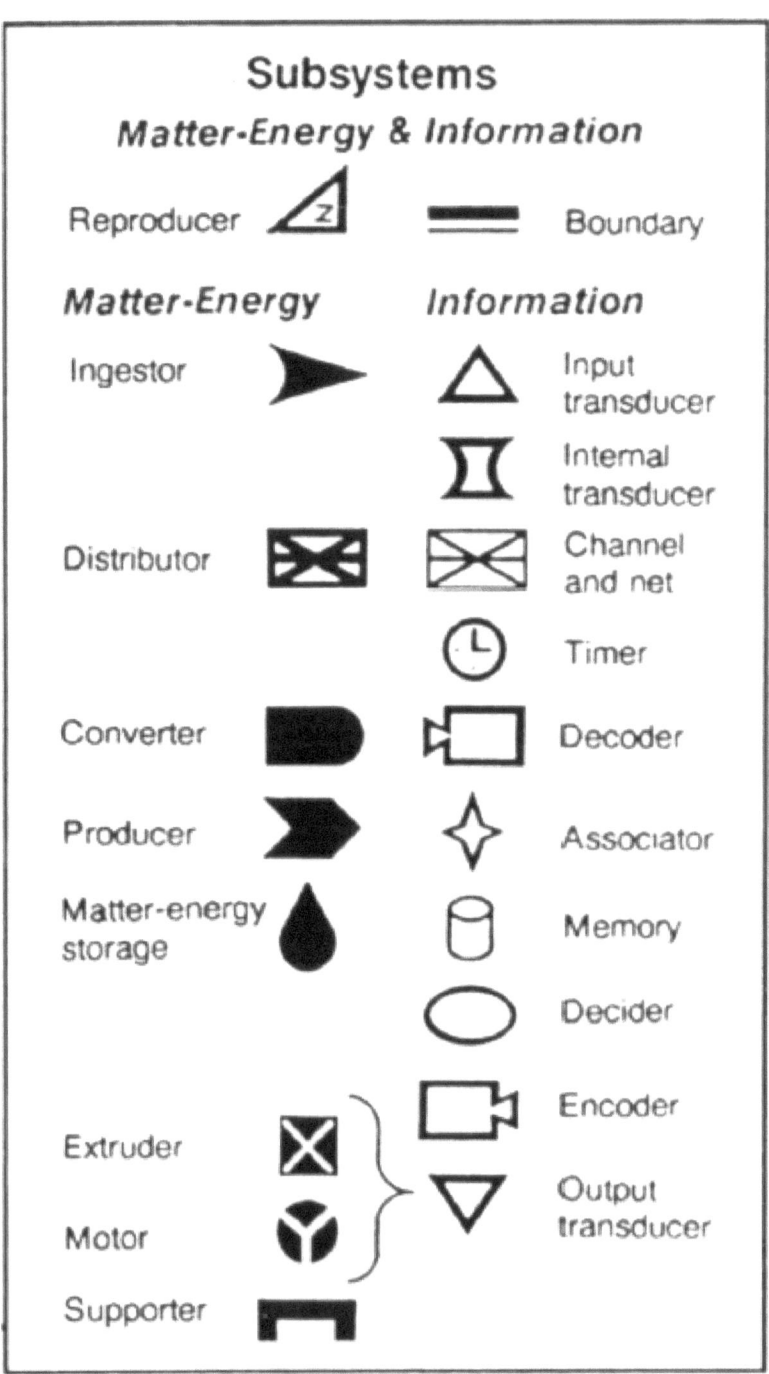

Subsystems
Matter-Energy & Information

Reproducer — Boundary

Matter-Energy   Information

Ingestor   Input transducer

Internal transducer

Distributor   Channel and net

Timer

Converter   Decoder

Producer   Associator

Matter-energy storage   Memory

Decider

Encoder

Extruder

Output transducer

Motor

Supporter

Such an analysis would take into account the primary needs of human systems - foraging for food and other necessary forms of matter and energy: feeding; fighting against environmental threats and stresses; fleeing from environmental dangers; and, in organizations which provide a comfortable, long-term habitat, perhaps reproducing the species. This study would analyze the effects on human social and individual behavior of such factors as weightlessness or 116 gravity; limited oxygen and water supplies; extreme temperatures; available light, heat, and power; varying patterns of light and dark; and so forth. A data bank or handbook could be developed of the values of multiple variables in each of the 20 subsystems of such a social system.

Economic, legal, political, international, and environmental aspects of a large and diverse space program using nonterrestrial resources require careful consideration.

The costs and benefits (both economic and nonfinancial) of programs utilizing nonterrestrial materials in space must be carefully analyzed. Detailed parametric models are needed which can be periodically updated as new data become available. Innovative means for financing such programs need to be found. Perhaps legislative initiatives should be taken to strengthen NASA's autonomy and enable the agency to enter into joint ventures with the private sector both here and abroad. We recommend continued exploration of new means for increasing nongovernmental participation in the space program, both to spread risks and costs and to broaden the advocacy base for a large space program benefiting from the use of resources off the Earth. Insurance for risk management and investment strategies for up-front capitalization should be examined.

We recommend exploration of ways to serve American national interests through either cooperative or competitive activities involving other nations in space. The relationship of the use of space resources to existing space treaties should be carefully examined.

Conclusion

It is our consensus that, after the space station becomes operational, any of several driving forces will result in an American initiative beyond LEO and CEO. That initiative might take any of several forms, but every scenario that we considered involves some combination of automated and human activities on the Moon. If a manned return to the Moon is chosen as a goal, then the prospect (and even the necessity) of using local resources arises. In this study we have examined some of the prospects for doing that, and we have recommended advance preparations toward the goal. These advance preparations are, in our judgment, practical and rewarding in proportion to their cost. We have identified places in existing Government organizations and programs where they could be carried out.

Because the Moon is believed to be deficient in some of the needed resources while meteorites are known to contain them, we have also recommended expansion and exploration of the known population of near-Earth asteroids, so that the role of this natural resource in space programs of the future can be properly evaluated. Also, we have noted the evidence that Mars and its satellites can provide local resources for missions there. We have not tried to predict just which objectives future Governments may aim toward. Instead, we have endeavored to define the nearer term technology measures that will be needed in any case and the nearer term flight projects which, if carried Out, would broaden our understanding of the natural resources available in space.

Recommendations

Our main recommendations (unranked) are as follows:

- Include growth provisions in current space station and orbital transfer vehicle systems to enable them to evolve into a cislunar infrastructure.
- Conduct laboratory research and development on a variety of ways to process lunar and meteoritic materials and make useful products from them.
- Support planetary observer missions with objectives of gathering scientific information and exploring resources. Such missions might include
    - Lunar polar geochemical orbiter
    - Mars (and martian satellite) observer
    - Multiple near-Earth asteroid rendezvous
- Discover and characterize more near-Earth asteroids by Earth-based telescopic observations.
- Develop advanced propulsion technology to permit comparison and selection of systems for transport beyond Earth orbit.
- Continue closed-ecosystem research and development with the aim of reducing resupply transport demand.
- Expand research on the challenges of living off Earth, including habitat design, space ecology, human/machine interactions, human-rating of equipment, and human behavior in remote sites—physiological, psychological, and social.
- Emphasize physical experiments and hardware development in preference to more paper studies.

Not only should NASA increase funding in these areas, but also other funding sources, both public and private, should be explored for possible support of the recommended research. University and industrial foundations, private institutions such as the Space Studies Institute and the Planetary Society, and new entities such as space business enterprises all have sponsored small research efforts related to their interests and might do so in this case—if, and only if, the future importance of nonterrestrial resources is made credible.

We have, we believe, sketched a coherent program of activities, engaging the talents of Government, industry, and the research community, with an easily supportable initial funding level, that could gather essential knowledge and build advocacy for the day when Americans will once more bravely and confidently set out on voyages of discovery and settlement—this time to the Moon and beyond.

Addendum: Participants The managers of the 1984 summer study were David S. McKay, Summer Study Co-Director and Workshop Manager Lyndon B. Johnson Space Center Stewart Nozette, Summer Study Co-Director California Space Institute James Arnold, Director of the California Space Institute Stanley R. Sadin, Summer Study Sponsor for the Office of Aeronautics and Space Technology NASA Headquarters Those who participated in the 10-week summer study as faculty fellows were the following: James D. Burke James L. Carter David R. Criswell Carolyn Dry Rocco Fazzolare Tom W. Fogwell Michael J. Gaffey Nathan C. Goldman Philip A. Harris Karl R. Johansson Elbert A. King Jesa Kreiner John S. Lewis Robert H. Lewis William Lewis James Grier Miller Sankar Sastri Michele Small Jet Propulsion Laboratory University of Texas, Dallas California Space Institute Virginia Polytechnic Institute University of Arizona Texas A & M University Rensselaer Polytechnic Institute University of Texas, Austin California Space Institute North Texas State University University of Houston, University Park California State University, Fullerton University of Arizona Washington University, St. Louis Clemson University University of California, Los Angeles New York City Technical College California Space Institute Participants in the 1-week workshops included the following: Constance F. Acton Bechtel Power Corp. William N. Agosto Lunar Industries, Inc. A. Edward Bence Exxon Mineral Company Edward Bock General Dynamics David F. Bowersox Los Alamos National Laboratory Henry W. Brandhorst, Jr. NASA Lewis Research Center David Buden NASA Headquarters Edmund J. Conway NASA Langley Research Center Gene Corley Portland Cement Association Hubert Davis Eagle Engineering Michael B. Duke NASA Johnson Space Center Charles H. Eldred NASA Langley Research Center Greg Fawkes Pegasus Software Ben R. Finney University of Hawaii Philip W. Garrison Jet Propulsion Laboratory Richard E. Gertsch Colorado School of Mines Mark Giampapa University of Arizona Charles E. Glass University of Arizona Charles L. Gould Rockwell International Joel S. Greenberg Princeton Synergetics, Inc. Larry A. Haskin Washington University, St. Louis Abe Hertzberg University of Washington Walter J. Hickel Yukon Pacific Christian W. Knudsen Carbotek, Inc. Eugene Konecci University of Texas, Austin George Kozmetsky University of Texas, Austin John Landis Stone & Webster Engineering Corp. T. D. Lin Construction Technology Laboratories John M. Logsdon George Washington University Ronald Maehl RCA Asiro-Electronics Thomas T. Meek Los Alamos National Laboratory Wendell W. Mendell NASA Johnson Space Center George Mueller Consultant Kathleen J. Murphy Consultant Barney B. Roberts NASA Johnson Space Center Sanders D. Rosenberg Aerojet TechSystems Company Robert Salkeld Consultant Donald R. Saxton NASA Marshall Space Flight Center James M. Shoji Rockwell International Michael C. Simon General Dynamics William R. Snow Electromagnetic Launch Research, Inc. Robert L. Staehle Jet Propulsion Laboratory Frank W. Stephenson, Jr. NASA Headquarters Wolfgang Steurer Jet Propulsion Laboratory Richard Tangum University of Texas, San Antonio Mead Treadwell Yukon Pacific Terry Triffet University of Arizona J. Peter Vajk Consultant Jesco von Puttkamer NASA Headquarters Scott Webster Orbital Systems Company

Gordon R. Woodcock Boeing Aerospace Company The following people participated in the summer study as guest speakers and consultants: Edwin E. "Buzz" Aldrin Research & Engineering Consultants Rudi Beichel Aerojet TechSystems Company David G. Brin California Space Institute Joseph A. Carroll California Space Institute Manuel I. Cruz Jet Propulsion Laboratory Andrew H. Cutler California Space Institute Christopher England Engineering Research Group Edward A. Gabns NASA Headquarters Peter Hammerling LaJolla Institute Eleanor F. Helin Jet Propulsion Laboratory Nicholas Johnson Teledyne Brown Engineering Joseph P. Kerwin NASA Johnson Space Center Joseph P. Loftus NASA Johnson Space Center Budd Love Consultant John J. Martin NASA Headquarters John Meson Defense Advanced Research Projects Agency Tom Meyer Boulder Center for Science and Policy John C. Niehoff Science Applications International Tadahiko Okumura Shimizu Construction Company Thomas 0. Paine Consultant William L. Quaide NASA Headquarters Namika Raby University of California, San Diego Donald G. Rea Jet Propulsion Laboratory Gene Roddenberry Writer Harrison H. "Jack" Schmitt Consultant Richard Schubert NASA Headquarters Elie Shneour Biosystems Associates, Ltd. Martin Spence Shimizu Construction Company James B. Stephens Jet Propulsion Laboratory Pat Sumi San Diego Unified School District Robert Waldron Rockwell International Simon P. Worden Department of Defense William Wright Defense Advanced Research Projects Agency